See It Grow
CARROT

by Joyce Markovics

Consultant: Karen C. Hall, PhD
Botanist
Eugene, Oregon

BEARPORT
PUBLISHING

New York, New York

Credits

Cover, © Foxxy63/Shutterstock, © S.O.E/Shutterstock, and © Lotus Images/Shutterstock; Title Page, © Danny Smythe/Shutterstock; TOC, © Nattika/Shutterstock; 4–5, © S-F/Shutterstock; 4L, © BWFolsom/iStock; 4R, © Kobyakov; 6, © DiyanaDimitrova/iStock; 7, © sturti/iStock and ArtKolo/Shutterstock; 8, © ifong/Shutterstock, © Antonia Lorenzo/Shutterstock, and © Henri Koskinen/Shutterstock; 9, © Dwight Sipler/CC BY 4.0; 10, © Nigel Cattlin/Alamy; 11, © kviktor/Shutterstock; 11R, © Prostock-Studio/Shutterstock; 12T, © Prostock-Studio/Shutterstock; 12B, © Dimitris66/iStock; 13, © Krzysztof Slusarczyk/Shutterstock; 14–15, © ifong/Shutterstock, © Grant Heilman Photography/Alamy, © amophoto_au/Shutterstock, and © suthas ongsiri/Shutterstock; 16, © Annaev/Shutterstock; 17, © inga spence/Alamy; 18T, © pixitive/iStock; 18B, © jeffbergen/iStock; 19, © DougSchneiderPhoto/iStock; 20 (T to B), © OlegDoroshin/Shutterstock, © Christian Fischer/CC BY-SA 4.0, and © Sarefo/GFDL; 21, © Mikhail Abramov/Shutterstock; 22, © Keith Foster/Alamy; 23 (T to B), © Varts/Shutterstock, © Cary Bates/Shutterstock, © Mostovyi Sergii Igorevich/Shutterstock, and © Prostock-Studio/Shutterstock; Back Cover, © Suslik1983/Shutterstock.

Publisher: Kenn Goin
Senior Editor: Joyce Tavolacci
Creative Director: Spencer Brinker
Design: Debrah Kaiser
Photo Researcher: Thomas Persano

Library of Congress Cataloging-in-Publication Data

Names: Markovics, Joyce, author.
Title: Carrot / by Joyce Markovics.
Description: New York, New York : Bearport Publishing Company, 2019. | Series: See it grow | Includes bibliographical references and index.
Identifiers: LCCN 2018009293 (print) | LCCN 2018011405 (ebook) | ISBN 9781684027187 (ebook) | ISBN 9781684026722 (library)
Subjects: LCSH: Carrots—Juvenile literature.
Classification: LCC SB351.C3 (ebook) | LCC SB351.C3 M37 2018 (print) | DDC 635/.13—dc23
LC record available at https://lccn.loc.gov/2018009293

Copyright © 2019 Bearport Publishing Company, Inc. All rights reserved. No part of this publication may be reproduced in whole or in part, stored in any retrieval system, or transmitted in any form or by any means, electronic, mechanical, photocopying, recording, or otherwise, without written permission from the publisher.

For more information, write to Bearport Publishing Company, Inc., 45 West 21st Street, Suite 3B, New York, New York 10010. Printed in the United States of America.

10 9 8 7 6 5 4 3 2 1

Contents

Carrot 4

Carrot Facts 22

Glossary 23

Index 24

Read More 24

Learn More Online 24

About the Author 24

Carrot

Carrots are delicious.

They are often long and orange.

How did they get that way?

An American will eat about 10,000 carrots in his or her lifetime.

A carrot starts out as a tiny seed.

carrot seeds

Carrot seeds are very small. A teaspoon can hold hundreds of them!

In early spring, the seed is planted in the ground.

Then, the seed is watered.

All plants need sunlight and water to grow.

About ten days later, the seed begins to grow.

shoot

taproot

A tiny **taproot** stretches into the soil.

Then, a green **shoot** pokes up out of the ground.

8

seed leaves

The carrot's first two leaves are called seed leaves.

9

As the days pass, the tiny plant gets bigger.

Now it grows feathery leaves.

The taproot gets longer and thicker.

Soon, it will become a carrot.

taproot

leaves

Leaves use the sun's light to make food for the plant. This helps the plant to grow.

11

taproot

After a few weeks, the taproot turns orange.

It has two layers—the phloem and the xylem.

These layers help carry food and water throughout the plant.

phloem

xylem

Carrots can also be purple, white, or yellow!

13

The taproot grows and grows.

14

leg

If it hits a rock, it might form extra roots called "legs."

Or it might grow sideways!

Carrots are **root vegetables**.

After about 75 days, the carrot is ready to be picked.

A farmer pulls it from the soil.

People grow carrots in gardens or on farms.

17

Twist off the carrot's leafy green top.

Wash it.

Peel it.

Then, bite into the tasty root.

Crunch, crunch, crunch!

Carrots can be eaten raw or cooked.

What happens to a carrot that's left in the ground?

Next spring, it will **produce** flowers.

The flowers become seeds.

Carrot flowers are tiny and white.

The seeds grow into new carrot plants!

Carrot Facts

- The longest carrot ever grown was nearly 17 feet (5 m) long!

- The heaviest carrot ever grown weighed almost 19 pounds (8.6 kg).

- Carrots are 87 percent water.

- China grows the most carrots of any country.

- Eating too many carrots can cause your skin to turn yellow!

Glossary

produce (pruh-DOOSS) to make

root vegetables (ROOT VEJ-tuh-buhlz) parts of plants that grow underground that people eat

shoot (SHOOT) a young plant that has just appeared above the soil

taproot (TAP-root) the large main root of a plant

Index

flowers 20
leaves 9, 10–11, 18
roots 8, 11, 12, 14–15, 19
seeds 6–7, 8, 20–21
shoot 8
soil 8, 16
sun 7, 11
taproot 8, 11, 12, 14–15, 19
water 7, 22

Read More

Rice, Dona Herweck. *How Plants Grow (Time for Kids).* Huntington Beach, CA: Teacher Created Materials (2012).

Schuh, Mari. *Carrots Grow Underground (How Fruits and Vegetables Grow).* Mankato, MN: Capstone (2011).

Learn More Online

To learn more about carrots, visit
www.bearportpublishing.com/SeeItGrow

About the Author

Joyce Markovics lives in Ossining, New York. She has a pet rabbit who loves to eat carrots—and bananas!